When the Nest Fell

Based on a True Animal Rescue Story

written by Kirsten Norton
illustrated by Jacqueline Gutierrez

Sprouting Seed Press
New York

When the Nest Fell
Based on a True Animal Rescue Story

written by Kirsten Norton, illustrated by Jacqueline Gutierrez

Published by:
Sprouting Seed Press
690 Saratoga Road, Suite 183 Burnt Hills, NY 12027
https://www.sproutingseedpress.com

ISBN-13: 978-0-9965073-3-2

LCCN: 2025909720

Lesson plan ideas are available on the Sprouting Seed Press website:

https://sproutingseedpress.com

Special discounts on bulk quantities of Sprouting Seed Press books are available to schools, healthcare organizations, and other qualifying groups.

For details, please contact:

Group Sales Department
Sprouting Seed Press
690 Saratoga Road, Suite 183
Burnt Hills, NY 12027
http://www.sproutingseedpress.com
sales@sproutingseedpress.com

To my husband Ryan for always believing in me and never questioning it when I bring wild animals home - and to my children Hudson and Hazel. Thank you for showing me how wonderful life can be.

Nestled in the branches of a lofty tree lived four little squirrels and their mother. The babies didn't know much about the world yet, but they knew the familiar comfort of their warm nest, the love of their mother, and the sounds of nature around them. The whooshing wind lulled them to sleep each night. The babbling stream comforted them. They hummed along to the song of the cicadas, buzzing lazily in the summer sun.

One morning after the babies finished their breakfast, things sounded different. The animals in the nearby woods stood quiet and still.

Voices boomed, startling the squirrels. Bang! A truck door slammed below. A woman with a kind smile came out of her house and greeted the workmen, pointing to the branch where the babies slept.

Mama squirrel knew that she had to move her babies. She had no time to lose. She scampered down the tree to find another place to build their new nest.

As the loud and unfamiliar noises were getting closer, the babies trembled. The workers' machines made loud whirrs and buzzes that didn't sound like the familiar humming of cicadas. The workers' equipment roared as the branch shook.

The babies tumbled from their nest.

Thud!

The babies squeaked for help, but mama was too far away to hear. The men didn't notice what happened. They were focused on their work and had ear plugs in to protect their ears.

The babies snuggled together, too stunned to move.

When the workers finished cutting down the tree, they began to clean up. A young man was clearing branches and stopped suddenly when he saw the babies. "Hey! I think we cut down a nest!" he shouted. The other men rolled their eyes and kept working.

Watching from her window, the kind lady heard what the young man said. She ran over to help him. "What should we do?" She panicked.

Blood and bruises developed on the little squirrels' skin. She scooped up the babies and rushed them inside, away from the chaos.

"We need to call someone for help" she muttered to herself. She placed the nest in a cardboard box and rushed to her computer to search for wildlife rehabilitators in the area.

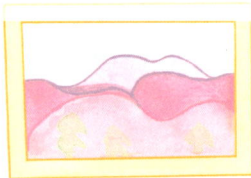

Meanwhile, mama squirrel watched the events unfold from the new nest she had created, but didn't see where her babies went. She promised herself she wouldn't stop looking until she found them. Unsure of where they had been taken, she jumped from tree branch to tree branch to search as far as her eyes could see.

The babies were getting quite hungry and thirsty. They hadn't had a drink since sunrise! They were achy from their fall and growing tired and weak. The kind lady covered them with a blanket and spoke softly. "Don't worry, we'll make this right," she said.

In the wild, baby squirrels nurse from their mother every 2 hours.

Thankfully, a rehabber finally knocked on the door.

The babies awoke from a nap to see a new face peering down at them. She smiled. "Hi there! I'm Kirsten. Let's get you fixed up" Kirsten lifted the squirrels and placed them into a heated container. She gently inspected the squirrels to determine their injuries, and reassured them they'd be okay. "You poor things, I'll let you rest before we treat your wounds at my rehab center."

When they arrived at Kirsten's rehab, the squirrels peered out and saw other creatures that were receiving treatment. Feeling less alone, they drifted into a deep sleep, even though they really missed their mama.

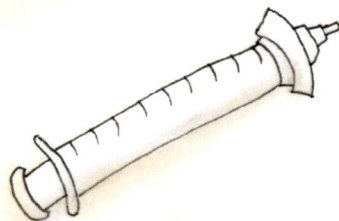

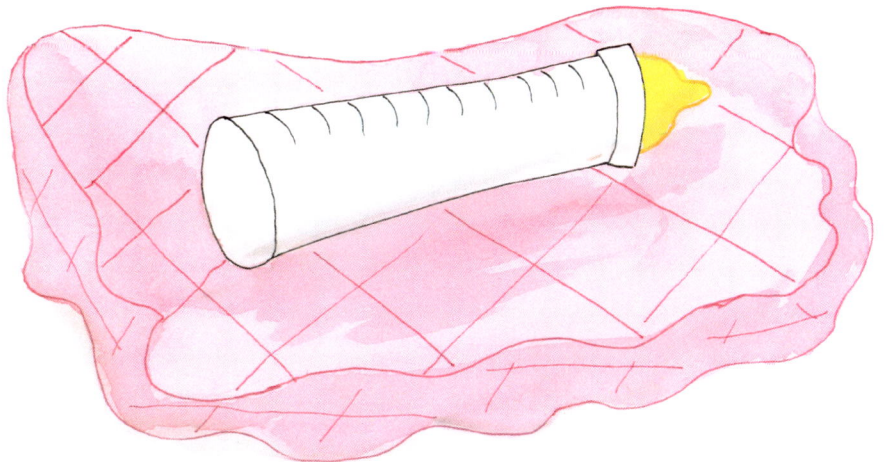

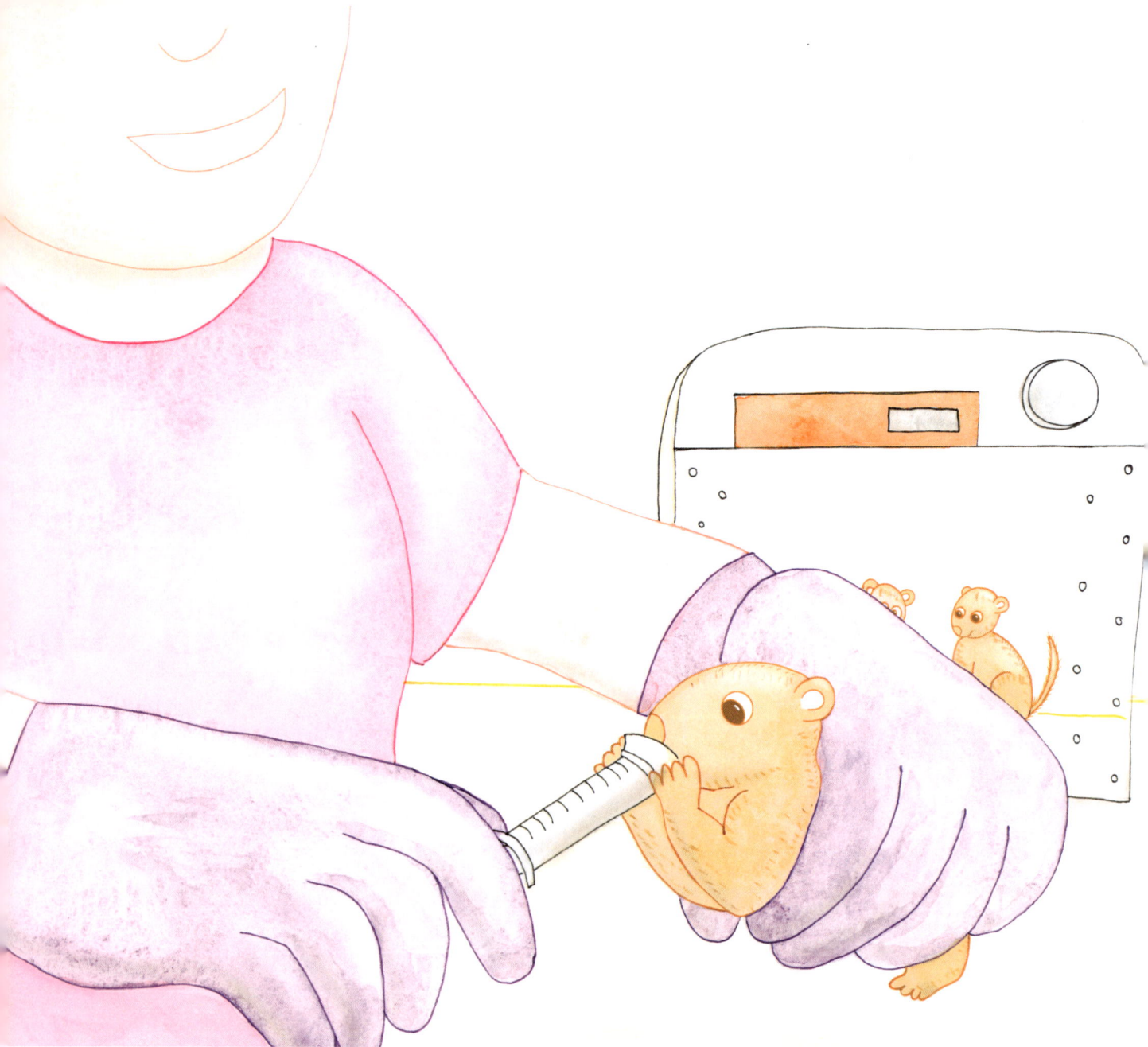

After the babies rested and warmed up, Kirsten began cleaning their wounds and gave them an electrolyte drink. "I'm going to take care of you, from tip to tail" Kirsten said, as the babies gulped down formula.

She came back to care for them every two hours through the night. "Your mama must be looking for you. We'll go out to find her in the morning" she said.

With each feeding, the squirrels became stronger. Their bodies felt better from the sleep and hydration, and their wounds felt less painful after being cleaned.

As with all animals that she takes in, Kirsten named the squirrels.

Kirsten drove them back to the site where they lost their home and their mother.

Cherry squeaked and twitched her tail in excitement when she saw their section of the woods. Their old nest was placed on top of the stump of their tree, now only a foot off the ground.

While they waited for their mama, Kirsten counted the rings of the tree to see how old it was. She felt sad about their tree, but didn't have much time to dwell on that.

She counted 120 rings, so it was around 120 years old!

Their mama returned when she heard their cries and she picked them up one by one to join her in their new nest.

When all the babies had been reunited with their mother, they snuggled together to mark each other with their scent. With each inhale and exhale they felt relieved to be back home again.

Mama looked at Kirsten as if to say thank you.

Kirsten smiled at her and said to the squirrels, "You must have been wondering why you were taken away and what happened to your tree. Unhealthy trees need to be cut down to prevent them from falling on people's homes. Most people don't realize that trees should be cut down in winter to prevent hurting wildlife nesting in the trees. I know that the kind lady wanted to help. She took you inside away from danger. I will explain to her that your mama would've been able to find you if you were left near the tree stump."

Kirsten waved goodbye and went to find the lady who owned the house.

As Kirsten spoke with the kind lady, mama squirrel curled beside her babies, who were safe in the new nest at last. They drifted off to sleep.

The cicadas buzzed and the wind blew softly. Everything was right in their world again.

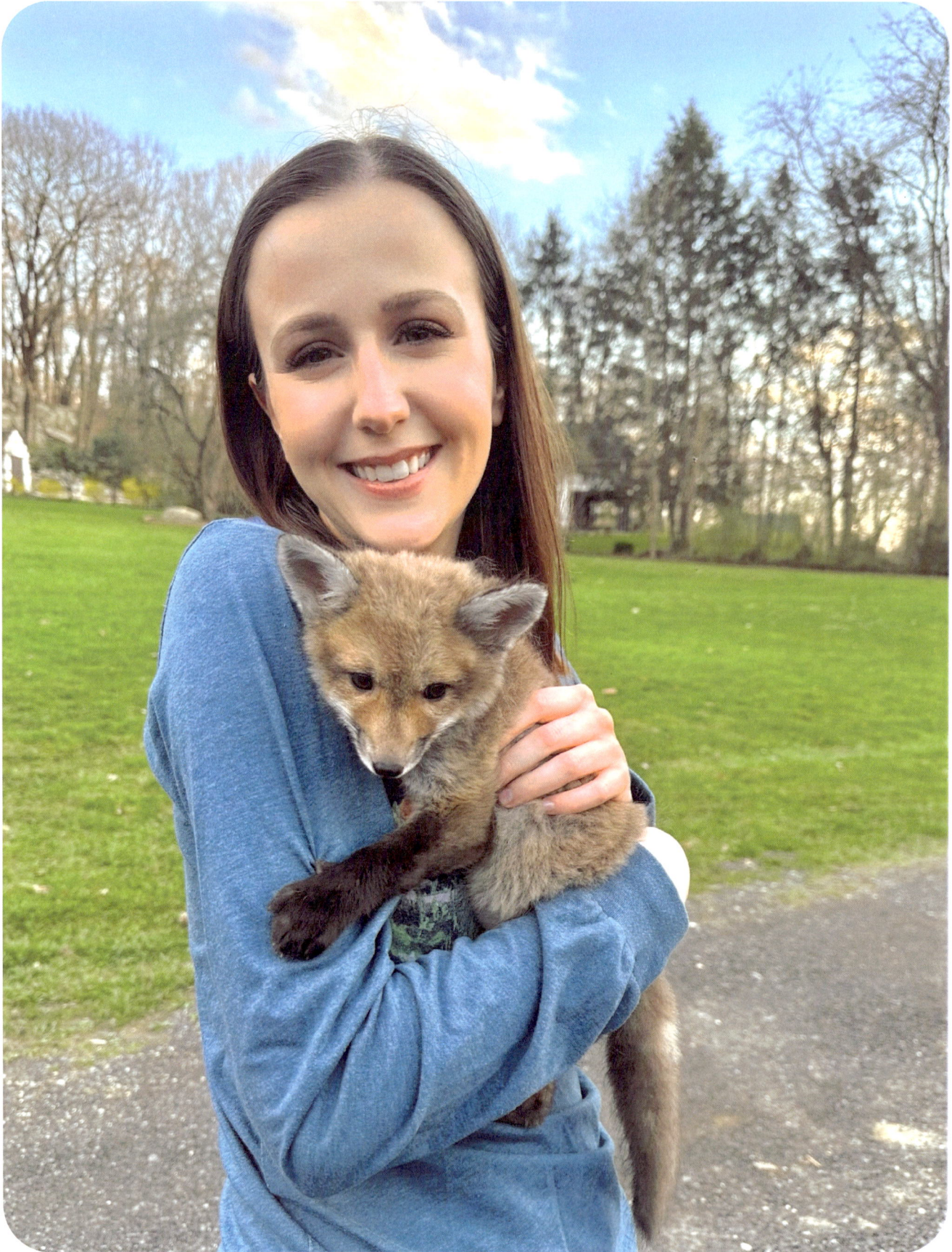

Behind the Scenes of Wildlife Rehabilitation

My name is Kirsten and I am a licensed New York State Wildlife Rehabilitator at The Injured Reserve, https://theinjuredreserve.com. This means I help animals! I wrote this story based on true events that I experienced while helping the squirrels in this book.

Like many animals that come into my rehab, these squirrels needed care before they could be released back into the wild again. When I got the call from the woman mentioned in this book, I wasn't sure if I should get involved. Often, when baby squirrels are separated from their mother, the best thing to do is to allow mom to help her babies. Young wildlife has the best chance of surviving when they are with their mother, even when they are a little injured. In this case, I decided they needed help because they had injuries that put them at risk for infection, and because they had been separated from their mom for so long that they were dehydrated. Baby squirrels normally feed every two hours and sleep for 22 hours a day.

The excitement from their nest falling also put them into shock, which means they were very afraid and couldn't calm down. When I took them to my home, where the Injured Reserve is located, I immediately got them warm and left them in a quiet room to rest. It's never a good idea to force feed wildlife when they are in shock. Giving them a quiet place to calm down is so important. Once these young squirrels were rested, I gave them plain Pedialyte to replace minerals that they lost from going so long without drinking. I also gave them each a body exam to find any injuries and make a treatment plan. Once hydrated, warm, and calm, I gave them formula specially made for baby squirrels. This is another reason to bring injured wildlife to rehabbers---we have all of the tools needed to help the animals! They shouldn't drink regular milk or human formula.

Then, I treated their injuries and left them to sleep. I repeated this process every two hours until the next morning when I went out at dawn and placed their drey on the stump of their tree. I waited a while and watched from a distance. I quietly left after I saw mom return and pick up one of her babies. I returned in an hour and saw she had taken all but one baby back. I was a little nervous because the one left was the most injured. I worried that mom might not accept this one because of its injuries. I was prepared to take this baby back and raise him to adulthood before releasing him to the wild. Thankfully mom did come back for him and all of them made it to their new nest.

This story is an example of a day in the life of what wildlife rehabilitators do. In New York, this is a volunteer position and people choose to do this because they love animals and the environment. We take in injured, orphaned and sick wildlife that are native to the state of New York that need care before they can be sent back to live in the wild. Each state in the U.S. has wildlife rehabilitators who help injured animals in their area.

There are lots of rules that we must follow to keep our licenses. It takes a long time to learn how to help animals in different situations. Every day I am still learning and I rely on more experienced rehabbers to give advice in situations that are new to me. I am sometimes scared to help, especially when I am unsure if the animal will survive. But, the best thing I can do is to be brave and give it my best try. The first step in helping any animal is getting them into a safe, quiet and calm environment so they can take a deep breath and calm down. Once calm, we always ensure they are hydrated before we begin treating them, like we did with the squirrels in this story.

If you find injured, orphaned or sick wildlife please call your local wildlife rehabilitator. You can do a quick Google search for "wildlife rehabilitators near me", visit your state's Department of Environmental Conservation/Protection website, (https://dec.ny.gov/nature/wildlife-health/rehabilitators in NY) or use the Animal Help Now app (available on Apple and Android) to locate a rehabber. Before intervening, ask yourself if the animal really needs help. Many animals who may appear to be orphaned are not abandoned. The mother is usually nearby but out of sight. A rehabber can give advice on how to tell if a baby animal really is in need of help. In any situation, it is best to speak with a professional first and it is always helpful to have as many details and pictures as possible, especially if you are unable to stay close to the animal in need.

Glossary

Achy - To be very sore and tired all over

Booming - A very loud and deep sound

Cicada - An insect that rubs its wings together to make a buzzing sound, usually heard in the summer

Dehydrated - To be very thirsty due to a loss of fluids in the body

Devastated - Very shocked and sad

Drey - A squirrel nest, found in the top of a tree and made with twigs

Ear plugs - These are used to block out noise. They plug the ears to keep noises out

Equipment - Tools needed to perform a task

The Injured Reserve - The name of Kirsten's wildlife rehab center

Interfere - To get involved with something you shouldn't

License - Formal, written permission from an organization permitting someone to do something that is not otherwise allowed

Lofty - Very tall and high up

Nature - The natural world, all the parts of the world that are not man made

Nest - A home made of branches and leaves for squirrels or other animals

Nestled - To cuddle up

Nurse - To take care of or to drink milk

Opportunity - A chance

Orphaned - A baby or child who lost their parents

Professional - Someone who does a job that they received training for and have experience in

Reassure - To comfort and soothe someone to help their fears go away

Scamper - To run around playfully

Startled - To feel surprised, shocked, and sometimes scared

Tree rings - Circular rings that are shown when a tree is cut open to show a cross sectional view. Each ring is left when the tree grows each year. This is how we tell how old a tree is.

Tremble - To shake in fear

Whirr - A buzzing noise

Wildlife - Wild animals and plants - anything alive in nature that was not domesticated by humans

Wildlife rehabilitators - Volunteers who work to help sick, injured and orphaned wildlife

Kirsten Norton is a New York State Licensed Wildlife Rehabilitator at the Injured Reserve in Suffolk County, NY. The Injured Reserve is a home-based animal rehab center that she runs independently. She has been helping animals since 2020 as a veterinarian technician and since 2022 as a wildlife rehabilitator. Her love of animals and her passion for storytelling inspired her to write this book, which is based on a true story. She originally intended the story to have an audience of one, as she wrote it for her son Hudson to enjoy. Encouraged by her husband, she decided to reach out to Sprouting Seed Press to publish the book. She lives in Suffolk County, New York with her family and hopes to inspire young readers to be stewards of the Earth.

Jacqueline Gutierrez is a writer, illustrator, dietitian, and surface pattern designer living in Saratoga County, NY. She writes and illustrates children's books, coloring books, and chapter books as well as educational articles related to health and nutrition for adults. Her illustrations are often inspired by science and nature, tell stories, and portray action in a visual format. She enjoys combining her experience as a dietitian, teacher, artist, and mother with her knowledge of art, science, and the natural world. She uses a combination of watercolors, gouache, pen and ink, and digital illustration to create her art.

www.ingramcontent.com/pod-product-compliance
Lightning Source LLC
Chambersburg PA
CBRC101308020426
42333CB00008B/78